The App Revolution: Be a Part of the Curve

By Melissa English & Bethany Collins

Preface:

Life is busy, harried, and often leave little time for much else. Have you ever thought, how much life would be easier with a certain application? In the world new applications are arising all of the time with an average of 109,500 being created a year. The right idea and the right design can land you a mint, but the most complicated thing is coming up with that amazing idea. How many different problems or issues can you imagine with a way to solve? Make a list and then enjoy this how to guide which will assist you with every step of the way creating your own app. Best of luck and happy coding, your million dollar idea!

-Ali Naguib

Table of Contents:

Chapters:

Introduction:

The world has been revolutionized and will continue to be so more and more as the years pass, currently technology is changing the way we do everything. This means that every year in time we advance we are advancing at least 10 years in technology. It is very difficult for people to catch up with the curve of technology and to learn how to use all of the new items which are coming out of development at an amazingly fast rate. And from this Apps were born, a way to make life a little easier and to help us handle all of the things which have become slightly complicated with the march of the beat of technology. There are apps for almost anything you can imagine and the apps which are out there are changing the way we live our lives and the way we do business in general. As everyone comments, there is an app for that. Anything that you can ask for you will find a way to do it through the use of an app.

And millionaires are being made every day with the development of more apps. Games like Angry Birds have even gone as far as having movies made about them. The range of creativity and imagination leaves an amazing spark of inspiration in the minds of all of the people who are using them. Each user is wondering how they can come up with an idea and make it their own. And this is your step by step guide to solve just that question. Join the curve and live your dreams!

The History:

In 2008, the iTunes App Store launched. Over 500,000 apps have been approved by Apple since then. Apple's IPhones are on the rise to becoming the most powerful mobile platform available through its reliance on desktop class operating system software. You don't have to be Stephen Hawking to write an app, either. These days, just about anyone with access to a computer can join the fun, even if you have never written a line of code!

So, how can a person with limited technical skills create an iPhone app? Anyone can make an one, it's just a matter of knowing the series of actions needed to make it happen. You will need an Apple computer with an Intel processor and Mac OS X Leopard, which is the sixth major release of Apple's desktop and server operating system for Macintosh computers. If you only have a PC, you will need to have at least a Dual Core processor and your motherboard/CPU combination must support VT-x (Hardware Virtualization), otherwise you will not be able to setup the Snow Leopard into Virtual Machine.

You can setup Sun Virtual Box as described here:
http://www.sysprobs.com/mac-os-guest-virtualbox-326-snow-leopard-1064-windows-7-32

You will also need a basic understanding of some key terminology. Programming is done with a group of programs called an Integrated Development Environment, or IDE. IDEs are made by hardware and software companies to allow programmers to make programs to run on their platforms. IDEs have a main program in them that connect other programs to them, thus making them integrated environments for development. The three most import programs in the IDE are the editor, the compiler and the simulator. An editor is any program that allows you to write and save text in a simple format, or plain text. Word processors such as Microsoft Word can be used as editors, but usually more specialized programs written specifically for IDEs are utilized. A compiler is a program that takes the code in the editor's text files and translates it into code that the computer can run. You use the main IDE program to tell the compiler the location of the files, called a bundle. A bundle is a folder that appears as a single file in the Finder. This convenience allows users to easily install and move applications without having to worry about keeping their support files around. A simulator, also known as an emulator, is a program that pretends to be a physical hardware device. For instance, Apple's free iPhone Software Development Kit is a terrific iPhone Simulator you can use to test and preview your websites. It takes the files produced by the compiler and runs them just like the original device. Apple's IDE, called Xcode, is available on Apple's software development website free to anyone registered as a developer. The iPhone SDK is repackaged Xcode with iPhone development resources.

Instructions for Registering as a Developer and Downloading Xcode:

1. Open your favorite web browser and go to http://developer.apple.com/iphone/program.

2. Scroll to the bottom of the page and click "Download the Free SDK".

3. Click "Create Apple ID".

4. Input your information and click "Create".

5. Click "iPhone SDK" in the downloads section.

6. Choose a location to save the DMG file.

Installing Xcode:

1. Double click the DMG and you will be presented with a

window containing the Xcode installation package.

2. Double click iPhone SDK.

3. Follow the installation instructions.

Running Xcode

 Xcode does not install into the Applications folder because it depends on many system-level services and includes many programs.

To run Xcode and other tools that come with the iPhone SDK:

1. Double-click "Macintosh HD" on your desktop.

2. Double-click "Developer"

3. Double-click "Applications"

4. Double-click "Xcode"

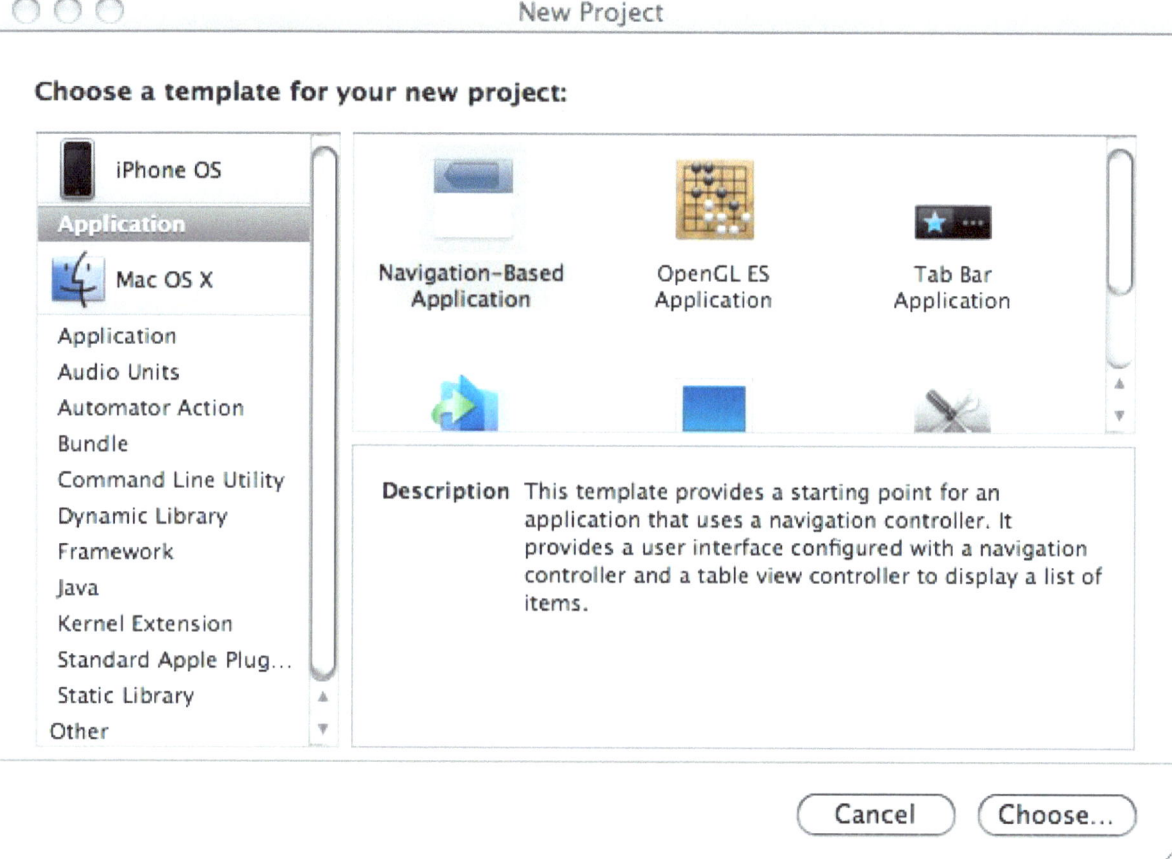

Creating Your First Project

Like most Mac applications, Xcode's user interface is composed of many windows connected

together with the main menu at the top. The first time Xcode is run, it will present a welcome

screen. These screens can be helpful, but close them for now and we will create your first iPhone

project.

1. Click "File -> New Project".

2. Click the "View-Based Application" icon.

3. Click "Choose".

4. Navigate to a location where you would like to store your iPhone projects. Name your project and save it the text box labeled "Save As:"

5. Click "Save".

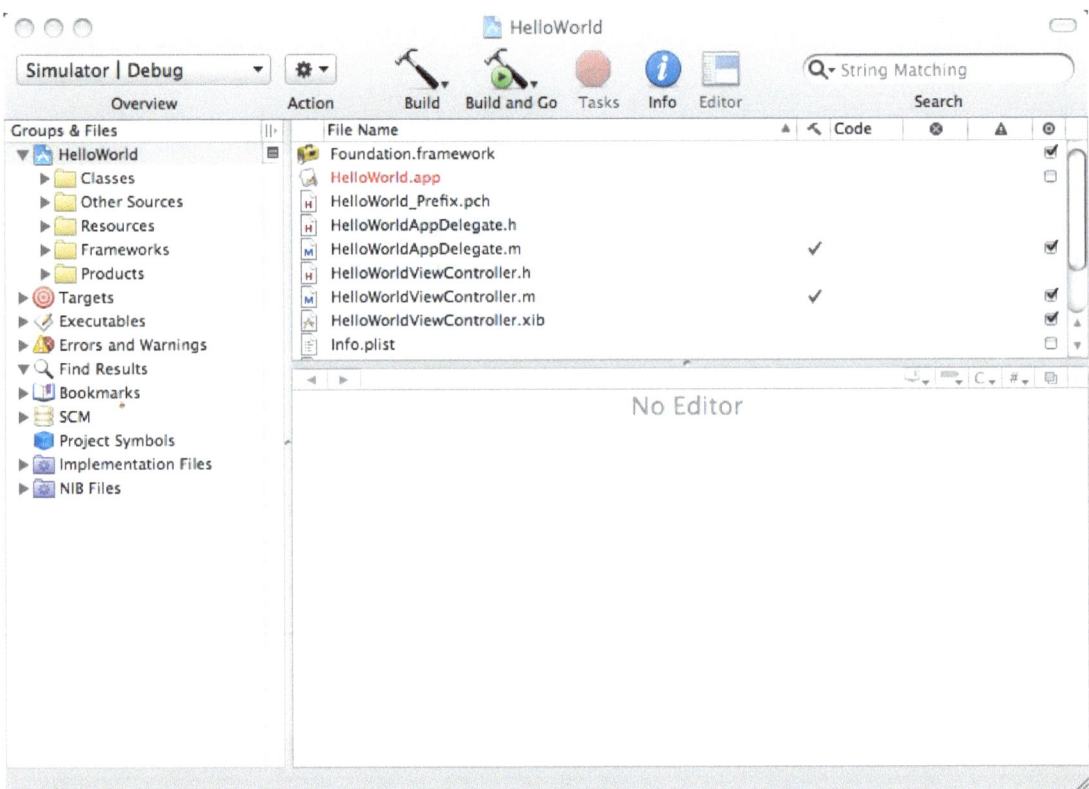

Writing and Compiling Your Program

The window you are seeing is the main IDE program that has the editor and controls for the compiler built into it. We will now add the code to create a button and a button action.

1. Expand the "Classes" group on the left side by clicking on the small triangle next to it.

2. Click on "HelloWorldViewController.m".

3. Scroll down to the second green code region. This is a code comment showing

where to put your code if you are creating your user interface using code.

4. Highlight the whole comment from the "/*" line to the "*/" line and paste or type

the following code in it's place:

- (void)viewDidLoad {[super viewDidLoad];UIButton* helloButton = [UIButton
buttonWithType:UIButtonTypeRoundedRect];helloButton.bounds = CGRectMake(0.0f, 0.0f,
200.0f, 50.0f);helloButton.center = self.view.center;helloButton.font = [UIFont
boldSystemFontOfSize:30.0f];[helloButtonsetTitle:@"Hello"forState:UIControlStateNormal];[h
elloButton addTarget:self action:@selector(helloButtonPressed:)
forControlEvents:UIControlEventTouchUpInside];

[self.view addSubview:helloButton];}- (void) helloButtonPressed:(id)sender {

[(UIButton*)sender setTitle:@"Hello World!" forState:UIControlStateNormal];}

Running Your Program

Now we have some code written for the compiler to compile and the simulator to run. Press the

"Build and Go" button at the top to compile your code and start the simulator running your

program. If asked to save your changes, press "Save All". The simulator will start up and run

your program. Congratulations on your first iPhone program!

Two years ago, when you could launch an app without much promotion and still have a

fair shot. Today, App Store competition is fierce and dozens of new high-quality apps are being

launched every day. This competitive environment means that ideas must be evaluated and

refined to achieve maximum success. Collaboration and evaluation are key to success in this venture. Seek out an expert with specialized knowledge of the industry. A variety of people are

qualified to evaluate ideas:

iPhone app marketing professionals-These people have specialized knowledge of how their

clients' apps have performed in the App Store, as well as promotional techniques.

iPhone app developers- Many successful indie developers are committed to serving their

community and are more than happy to help review your idea.

Personal connection- If you know someone personally who has created and launched an app,

take them out for lunch and ask for feedback.

Things to look for when getting expert opinions:
- ✓ Product competition
- ✓ Technical limitations
- ✓ Development process
- ✓ Increase sales opportunities
- ✓ Production cost estimates
- ✓ Mobile context fit
- ✓ Target audience feedback

Interactive apps with a niche that solve a unique problem are more successful. 80% of apps are

NOT generating enough revenue to support business. Here are some steps to take on your

journey:

Step 1: Develop A Monetization And Marketing Plan- figure out how your idea will generate a profit!

Step 2: Sign Up For A Developer Account- visit the iOS Development Center and sign up for an account. It's $99 for a year and requires that you provide the tax and bank account information of your business or yourself. (The only reason not to sign up for a developer account would be if you will accept having your app published under another person or company's account and brand. Apple would then pay all revenue to the account holder's bank account, who would then be responsible for paying you.)

Step 3: Sketch Your Application- start putting your ideas down on paper as to how the app would look, work and the information it would present. To sketch a rough interface, ask yourself:

What primary action will users take within the app?

What information will each screen need to present?

What is the flow? How will users get from start to finish?

How big should the elements on screen be relative to each other?

The right tools can simplify your sketching layout. Create at least one thumbnail sketch for each screen in your application. Experiment with various navigational schemes, the copy on buttons and the flow between screens. For transferring your sketches into digital format, iPlotz is a good tool. The purpose of sketching your application's screens is to build a foundation for the next phase of the project. If you're an entrepreneur and are outsourcing the bulk of the project's work, then you would show these sketches to the design and development team in order to get a price estimate.

Step 4: Identify Work Outsourced

You're starting a small business when you create an iPhone application. Can one person to play the role of researcher, project manager, accountant, information architect, designer, developer, marketer and advertiser. Of course, but they would be wasting a lot of time, energy and sanity in the process. The following areas define where you may need to hire help: design, programming, promotion and marketing. The least expensive way to produce your application would be by hiring freelance contractors. While your costs would be lower, your role as project manager would become more prominent. You would spend more time managing the project. Also, remember to have freelancers sign a contract with details on the scope of their work, your expectations and the payment terms. Another option is to hire an agency to handle the production. The agency would manage the project, and your role as client would be to review and approve its work. Working with an agency is a good option if you have a larger budget and less time to dedicate to the production process.

Step 5: Hiring A Team

Download my "iPhone App Template," a big collection of iPhone UI elements. These Photoshop files will save you a lot of time getting started on the design. To learn more about mobile design in general, these websites provide a lot of great resources:

Marc Edwards' articles

Ken Yarmosh's blog

Luke Wroblewski's blog

DesignBoost (disclaimer: this is my mobile design training website)

UI Stencils

If you're not a designer, then you should know that design breaks down into three roles: information architecture, interaction design and visual design. Finding one person with all of these skills is possible, but know that the design process calls for three distinct deliverables. If you've got your sketches, then you have everything you need for a designer to get started.

Information architecture- organizing the content in your app

Interaction design- the flow of the app from screen to screen

Visual design- the final step; it is the "skin" that overlays the controls for the app. The key is to focus on the usability and primary task of the app.

Try to find a designer who has experience designing for mobile devices. They will have some good feedback and suggestions to improve your sketches. A few places to look for designers:

Dribbble

crowdSPRING

Elance

Finding a developer

Using your sketches, compose a specification document that describes in writing what your app does. This document is what you will share with the developer to get a time and cost estimate. Having a document like this also ensures that you will be able to hire a developer who has the skills necessary to produce the app. Your developer can also help you submit your application to the App Store. Clearly communication between you and the developer is very important. Nothing is more detrimental to success than a poorly planned launch.

Here are a few places to look for developers:

oDesk

They Make Apps

Elance

Marketing and Promotion

You can have the best app out there and if it just sits on the shelf, it will never succeed. You must get your app exposure in order for it to be successful. There are many strategies for marketing and promotion. Incorporate social media and build that functionality into it. Set up fan pages for your app on Facebook and Twitter, and use them as platforms to communicate with users and get feedback. Start building buzz with pre-launch promotion about your app. Email journalists and bloggers who write about things related to your app. The more relevant your app is to their niche, the better your chances of getting written about. Plan for multiple releases by not packing your first release with every feature you want to offer. Periodically release new versions of the app to boost sales.

Try these:

TUAW (The Unofficial Apple Weblog)

Macworld

AppCraver

TouchArcade

Research was done to find out how companies with top-rated apps in markets like the Apple App Store or the Google Play Store design, build, test, release, and maintain their mobile apps. While there's great diversity in approach, common factors fell out of my interviews, and these practices will help you organize your own efforts:

1. Assemble small, focused development teams. The largest development teams found had fewer than 10 people. When teams grew larger, they were subdivided into platform-specific teams (i.e. the Android team or the iOS team). Small teams can move fast and keep impedance to a minimum but also require substantial changes in how they perform design and testing.

2. Favor simple development tools over complex ALM processes. The need for speed that characterized the development processes found means that most traditional ALM tools hurt more than they help. The emphasis is on smaller code bases with less branching mean tools instead of

formal, text-based requirements documents and heavyweight SCM systems. Experiment with device emulators and simulators to get early user feedback.

3. Balance release speed with a focus on quality. Teams that have honed their skills deploying in a web-centric world are in for a bit of a shock as they move to building mobile apps. In their web world, a focus on automation and attention to DevOps principles lower the impact of introducing regressions into a working system: you can always "roll forward" or redirect traffic from your blue environment back to the green one. It's not as easy with Apple, Microsoft, and Amazon standing between you and your customers with app store approval processes. The cost of failure and the time to recover from it put renewed emphasis on quality processes.

4. Prioritize gathering user feedback and forming a rapid response to it. Most of the changes seen with top-performing app development teams are based on the central premise of reducing the time to feedback. It's also important to monitor the customer feedback and organize the team to respond in real time. A response might be customer outreach, a request to rate an app, or restructuring the app backlog to prioritize a feature request. The public nature of the feedback process will throw many traditional development teams for a loop, but can work in your favor if done correctly.

Intuitive Design

The fewer registration buttons, the better. Impatient users will look for any excuse to back out of downloading an app. A great app is designed with a specific set of purposes in mind, and when you add unnecessary items, you dilute that functionality. To make people devoted to your app offer features no one else has, but make sure the design is flexible, customizable and intuitive. Every design element must be considered in terms of efficiency and functionality. Throughout this process new problems can be created where none previously existed so keep testing even the most minor functions. Obvious solutions might not seem so at the beginning of the process so don't be afraid to make last-minute tweaks or to let a trusted outsider test your product. Do try to anticipate potential usage problems and have customer-service options in place. Customer service is at the root of it all, so don't forget about the person you made your app for to begin with.

Nothing helps guide a design better than real world feedback. Take your design to your peers, your mother, your granddad, or even your future end users. You can even solicit feedback from a limited beta of your app or use various online solutions such as getsatisfaction.com. Ask your users what they like, what they don't like, how they would expect things to flow, and so on. Taking this feedback and integrating it into your app will help ensure users are happy. If you cannot easily gather end user feedback on the design and flow of your app, here are some other points to keep in mind at the beginning of your project:

- ❖ Keep the user interface and design simple. Your app should require a user to have little to no instructions in order to get started.

- ❖ Ensure proper spacing of objects for touch screen devices. Nothing will frustrate your users more if their touch events are misread.

- ❖ The layout of the screen should be well used, but not too cluttered. Space clean and evenly for objects in groups,

- ❖ Make use of menus when required. Menus are great when there are additional options or settings that you want a user to have access to, but don't want to have them always displayed on screen.

- ❖ Keep your app consistent. Every screen within the app should offer a similar user experience in layout, theme and flow. A user should know this is your app regardless of the screen they are looking at.

Tips and Tricks to Remember:

❖ Solve a problem that is causing issues for the world

❖ Keep your design consistent

❖ Remember not to share your app idea with anyone other than other co-creators

❖ Others might attempt to rush your idea to the market before you are able to.

❖ Be consistent in everything that you do.

❖ Find a great marketing representative.

❖ Use free media like You Tube, Mailchimp, and others to sell your product to the world as well.

❖ Have fun creating your app you are reaching out to the world and starting your own legacy.

About the Authors:

Melissa English

Melissa English is originally from Atlanta Georgia and loves to write fiction, nonfiction, screenplays, poetry and children's books. She was published for the first time at the age of 9, and has been writing ever since. She went to school at the University of Georgia receiving two masters degrees she has a Masters in Linguistics and Classics as well as a Masters in International Business. She works as a consultant and spends all the time she can writing. A few of her other titles *include My Truth, Molly McCreary: There's Something in the Garden, the Transiente Amores, Living Gluten Free on a Shoestring Budget, and The Amores, A Progression of the Rights of Love.* To learn more about Melissa English check out her website at Melissa English.net or you can email her at melissaenglish777@gmail.com for consulting purposes.

Bethany has her bachelor's degree in psychology from Warren Wilson College in Asheville North Carolina in completed graduate studies for reading education at the University of North Carolina at Asheville. She spent almost 10 years as an educator with him the public and private school arena. She currently lives in Atlanta Georgia and enjoy spending time with her dog. She has a popular blog about love, life, loss and transformation at http:// livelaughlove-beth.blogspot. com/ and is working on a memoir.

About the Publisher:

Vanity International Press is a Press located in Atlanta Georgia and owned by Melissa English. Please feel free to contact her with any questions!